ICME-13 Topical Surveys

Series editor

Gabriele Kaiser, Faculty of Education, University of Hamburg, Hamburg, Germany

More information about this series at http://www.springer.com/series/14352

Peter Liljedahl · Manuel Santos-Trigo
Uldarico Malaspina · Regina Bruder

Problem Solving
in Mathematics Education

Peter Liljedahl
Faculty of Education
Simon Fraser University
Burnaby, BC
Canada

Manuel Santos-Trigo
Mathematics Education Department
Cinvestav-IPN, Centre for Research
    and Advanced Studies
Mexico City
Mexico

Uldarico Malaspina
Pontificia Universidad Católica del Perú
Lima
Peru

Regina Bruder
Technical University Darmstadt
Darmstadt
Germany

ISSN 2366-5947   ISSN 2366-5955   (electronic)
ICME-13 Topical Surveys
ISBN 978-3-319-40729-6   ISBN 978-3-319-40730-2   (eBook)
DOI 10.1007/978-3-319-40730-2

Library of Congress Control Number: 2016942508

Printed on acid-free paper

This Springer imprint is published by Springer Nature
The registered company is Springer International Publishing AG Switzerland

Main Topics You Can Find in This "ICME-13 Topical Survey"

- Problem-solving research
- Problem-solving heuristics
- Creative problem solving
- Problems solving with technology
- Problem posing

Contents

Problem Solving in Mathematics Education

Mathematical problem solving has long been seen as an important aspect of mathematics, the teaching of mathematics, and the learning of mathematics. It has infused mathematics curricula around the world with calls for the teaching *of* problem solving as well as the teaching of mathematics *through* problem solving. And as such, it has been of interest to mathematics education researchers for as long as our field has existed. More relevant, mathematical problem solving has played a part in every ICME conference, from 1969 until the forthcoming meeting in Hamburg, wherein mathematical problem solving will reside most centrally within the work of Topic Study 19: Problem Solving in Mathematics Education. This booklet is being published on the occasion of this Topic Study Group.

To this end, we have assembled four summaries looking at four distinct, yet inter-related, dimensions of mathematical problem solving. The first summary, by Regina Bruder, is a nuanced look at heuristics for problem solving. This notion of heuristics is carried into Peter Liljedahl's summary, which looks specifically at a progression of heuristics leading towards more and more creative aspects of problem solving. This is followed by Luz Manuel Santos Trigo's summary introducing us to problem solving *in* and *with* digital technologies. The last summary, by Uldarico Malaspina Jurado, documents the rise of problem posing within the field of mathematics education in general and the problem solving literature in particular.

Each of these summaries references in some critical and central fashion the works of George Pólya or Alan Schoenfeld. To the initiated researchers, this is no surprise. The seminal work of these researchers lie at the roots of mathematical problem solving. What is interesting, though, is the diverse ways in which each of the four aforementioned contributions draw on, and position, these works so as to fit into the larger scheme of their respective summaries. This speaks to not only the depth and breadth of these influential works, but also the diversity with which they can be interpreted and utilized in extending our thinking about problem solving.

© The Author(s) 2016
P. Liljedahl et al., *Problem Solving in Mathematics Education*,
ICME-13 Topical Surveys, DOI 10.1007/978-3-319-40730-2_1

Taken together, what follows is a topical survey of ideas representing the diversity of views and tensions inherent in a field of research that is both *a means to an end* and *an end onto itself* and is unanimously seen as central to the activities of mathematics.

1 Survey on the State-of-the-Art

1.1 Role of Heuristics for Problem Solving—Regina Bruder

The origin of the word *heuristic* dates back to the time of Archimedes and is said to have come out of one of the famous stories told about this great mathematician and inventor. The King of Syracuse asked Archimedes to check whether his new wreath was really made of pure gold. Archimedes struggled with this task and it was not until he was at the bathhouse that he came up with the solution. As he entered the tub he noticed that he had displaced a certain amount of water. Brilliant as he was, he transferred this insight to the issue with the wreath and knew he had solved the problem. According to the legend, he jumped out of the tub and ran from the bathhouse naked screaming, "Eureka, eureka!". *Eureka* and *heuristic* have the same root in the ancient Greek language and so it has been claimed that this is how the academic discipline of "heuristics" dealing with effective approaches to problem solving (so-called heurisms) was given its name. Pólya (1964) describes this discipline as follows:

> Heuristics deals with solving tasks. Its specific goals include highlighting in general terms the reasons for selecting those moments in a problem the examination of which could help us find a solution. (p. 5)

This discipline has grown, in part, from examining the approaches to certain problems more in detail and comparing them with each other in order to abstract similarities in approach, or so-called heurisms. Pólya (1949), but also, inter alia, Engel (1998), König (1984) and Sewerin (1979) have formulated such heurisms for mathematical problem tasks. The problem tasks examined by the authors mentioned are predominantly found in the area of talent programmes, that is, they often go back to mathematics competitions.

In 1983 Zimmermann provided an overview of heuristic approaches and tools in American literature which also offered suggestions for mathematics classes. In the German-speaking countries, an approach has established itself, going back to Sewerin (1979) and König (1984), which divides school-relevant heuristic procedures into heuristic tools, strategies and principles, see also Bruder and Collet (2011).

Below is a review of the conceptual background of heuristics, followed by a description of the effect mechanisms of heurisms in problem-solving processes.

1.1.1 Research Review on the Promotion of Problem Solving

In the 20th century, there has been an advancement of research on mathematical problem solving and findings about possibilities to promote problem solving with varying priorities (c.f. Pehkonen 1991). Based on a model by Pólya (1949), in a first phase of research on problem solving, particularly in the 1960s and the 1970s, a series of studies on problem-solving processes placing emphasis on the importance of heuristic strategies (heurisms) in problem solving has been carried out. It was assumed that teaching and learning heuristic strategies, principles and tools would provide students with an orientation in problem situations and that this could thus improve students' problem-solving abilities (c.f. for instance, Schoenfeld 1979). This approach, mostly researched within the scope of talent programmes for problem solving, was rather successful (c.f. for instance, Sewerin 1979). In the 1980s, requests for promotional opportunities in everyday teaching were given more and more consideration: *"problem solving must be the focus of school mathematics in the 1980s"* (NCTM 1980). For the teaching and learning of problem solving in regular mathematics classes, the current view according to which cognitive, heuristic aspects were paramount, was expanded by certain student-specific aspects, such as attitudes, emotions and self-regulated behaviour (c.f. Kretschmer 1983; Schoenfeld 1985, 1987, 1992). Kilpatrick (1985) divided the promotional approaches described in the literature into five methods which can also be combined with each other.

- *Osmosis*: action-oriented and implicit imparting of problem-solving techniques in a beneficial learning environment
- *Memorisation*: formation of special techniques for particular types of problem and of the relevant questioning when problem solving
- *Imitation*: acquisition of problem-solving abilities through imitation of an expert
- *Cooperation*: cooperative learning of problem-solving abilities in small groups
- *Reflection*: problem-solving abilities are acquired in an action-oriented manner and through reflection on approaches to problem solving.

Kilpatrick (1985) views as success when heuristic approaches are explained to students, clarified by means of examples and trained through the presentation of problems. The need of making students aware of heuristic approaches is by now largely accepted in didactic discussions. Differences in varying approaches to promoting problem-solving abilities rather refer to deciding which problem-solving strategies or heuristics are to imparted to students and in which way, and not whether these should be imparted at all or not.

1.1.2 Heurisms as an Expression of Mental Agility

The activity theory, particularly in its advancement by Lompscher (1975, 1985), offers a well-suited and manageable model to describe learning activities and differences between learners with regard to processes and outcomes in problem solving (c.f. Perels et al. 2005). Mental activity starts with a goal and the motive of a person to perform such activity. Lompscher divides actual mental activity into content and process. Whilst the content in mathematical problem-solving consists of certain concepts, connections and procedures, the process describes the psychological processes that occur when solving a problem. This course of action is described in Lompscher by various qualities, such as systematic planning, independence, accuracy, activity and agility. Along with differences in motivation and the availability of expertise, it appears that intuitive problem solvers possess a particularly high mental agility, at least with regard to certain contents areas.

According to Lompscher, "flexibility of thought" expresses itself

> ... by the capacity to change more or less easily from one aspect of viewing to another one or to embed one circumstance or component into different correlations, to understand the relativity of circumstances and statements. It allows to reverse relations, to more or less easily or quickly attune to new conditions of mental activity or to simultaneously mind several objects or aspects of a given activity (Lompscher 1975, p. 36).

These typical manifestations of mental agility can be focused on in problem solving by mathematical means and can be related to the heurisms known from the analyses of approaches by Pólya et al. (c.f. also Bruder 2000):

Reduction: Successful problem solvers will intuitively reduce a problem to its essentials in a sensible manner. To achieve such abstraction, they often use visualisation and structuring aids, such as informative figures, tables, solution graphs or even terms. These heuristic tools are also very well suited to document in retrospect the approach adopted by the intuitive problem solvers in a way that is comprehensible for all.

Reversibility: Successful problem solvers are able to reverse trains of thought or reproduce these in reverse. They will do this in appropriate situations automatically, for instance, when looking for a key they have mislaid. A corresponding general heuristic strategy is working in reverse.

Minding of aspects: Successful problem solvers will mind several aspects of a given problem at the same time or easily recognise any dependence on things and vary them in a targeted manner. Sometimes, this is also a matter of removing barriers in favour of an idea that appears to be sustainable, that is, by simply "hanging on" to a certain train of thought even against resistance. Corresponding heurisms are, for instance, the principle of invariance, the principle of symmetry (Engel 1998), the breaking down or complementing of geometric figures to calculate surface areas, or certain terms used in binomial formulas.

Change of aspects: Successful problem solvers will possibly change their assumptions, criteria or aspects minded in order to find a solution. Various aspects of a given problem will be considered intuitively or the problem be viewed from a different perspective, which will prevent "getting stuck" and allow for new insights and approaches. For instance, many elementary geometric propositions can also be proved in an elegant vectorial manner.

Transferring: Successful problem solvers will be able more easily than others to transfer a well-known procedure to another, sometimes even very different context. They recognise more easily the "framework" or pattern of a given task. Here, this is about own constructions of analogies and continual tracing back from the unknown to the known.

Intuitive, that is, untrained good problem solvers, are, however, often unable to access these flexibility qualities consciously. This is why they are also often unable to explain how they actually solved a given problem.

To be able to solve problems successfully, a certain mental agility is thus required. If this is less well pronounced in a certain area, learning how to solve problems means compensating by acquiring heurisms. In this case, insufficient mental agility is partly "offset" through the application of knowledge acquired by means of heurisms. Mathematical problem-solving competences are thus acquired through the promotion of manifestations of mental agility (reduction, reversibility, minding of aspects and change of aspects). This can be achieved by designing sub-actions of problem solving in connection with a (temporarily) conscious application of suitable heurisms. Empirical evidence for the success of the active principle of heurisms has been provided by Collet (2009).

Against such background, learning how to solve problems can be established as a long-term teaching and learning process which basically encompasses four phases (Bruder and Collet 2011):

1. Intuitive familiarisation with heuristic methods and techniques.
2. Making aware of special heurisms by means of prominent examples (explicit strategy acquisition).
3. Short conscious practice phase to use the newly acquired heurisms with differentiated task difficulties.
4. Expanding the context of the strategies applied.

In the first phase, students are familiarised with heurisms intuitively by means of targeted aid impulses and questions (what helped us solve this problem?) which in the following phase are substantiated on the basis of model tasks, are given names and are thus made aware of their existence. The third phase serves the purpose of a certain familiarisation with the new heurisms and the experience of competence through individualised practising at different requirement levels, including in the form of homework over longer periods. A fourth and delayed fourth phase aims at more flexibility through the transfer to other contents and contexts and the increasingly intuitive use of the newly acquired heurisms, so that students can enrich their own problem-solving models in a gradual manner. The second and third phases build upon each other in close chronological order, whilst the first phase should be used in class at all times.

All heurisms can basically be described in an action-oriented manner by means of asking the right questions. The way of asking questions can thus also establish a certain kind of personal relation. Even if the teacher presents and suggests the line of basic questions with a prototypical wording each time, students should always be given the opportunity to find "their" wording for the respective heurism and take a note of it for themselves. A possible key question for the use of a heuristic tool would be: How to illustrate and structure the problem or how to present it in a different way?

Unfortunately, for many students, applying heuristic approaches to problem solving will not ensue automatically but will require appropriate early and long-term promoting. The results of current studies, where promotion approaches to problem solving are connected with self-regulation and metacognitive aspects, demonstrate certain positive effects of such combination on students. This field of research includes, for instance, studies by Lester et al. (1989), Verschaffel et al. (1999), the studies on teaching method IMPROVE by Mevarech and Kramarski (1997, 2003) and also the evaluation of a teaching concept on learning how to solve problems by the gradual conscious acquisition of heurisms by Collet and Bruder (2008).

1.2 Creative Problem Solving—Peter Liljedahl

There is a tension between the aforementioned story of Archimedes and the heuristics presented in the previous section. Archimedes, when submersing himself in the tub and suddenly seeing the solution to his problem, wasn't relying on osmosis, memorisation, imitation, cooperation, or reflection (Kilpatrick 1985). He wasn't drawing on reduction, reversibility, minding of aspects, change of aspect, or transfer (Bruder 2000). Archimedes was stuck and it was only, in fact, through insight and sudden illumination that he managed to solve his problem. In short, Archimedes was faced with a problem that the aforementioned heuristics, and their kind, would not help him to solve.

According to some, such a scenario is the definition of a problem. For example, Resnick and Glaser (1976) define a problem as being something that you do not have the experience to solve. Mathematicians, in general, agree with this (Liljedahl 2008).

> Any problem in which you can see how to attack it by deliberate effort, is a routine problem, and cannot be an important discover. You must try and fail by deliberate efforts, and then rely on a sudden inspiration or intuition or if you prefer to call it luck. (Dan Kleitman, participant cited in Liljedahl 2008, p. 19).

Problems, then, are tasks that cannot be solved by direct effort and will require some creative insight to solve (Liljedahl 2008; Mason et al. 1982; Pólya 1965).

1.2.1 A History of Creativity in Mathematics Education

In 1902, the first half of what eventually came to be a 30 question survey was published in the pages of L'Enseignement Mathématique, the journal of the French

Mathematical Society. The authors, Édouard Claparède and Théodore Flournoy, were two Swiss psychologists who were deeply interested in the topics of mathematical discovery, creativity and invention. Their hope was that a widespread appeal to mathematicians at large would incite enough responses for them to begin to formulate some theories about this topic. The first half of the survey centered on the reasons for becoming a mathematician (family history, educational influences, social environment, etc.), attitudes about everyday life, and hobbies. This was eventually followed, in 1904, by the publication of the second half of the survey pertaining, in particular, to mental images during periods of creative work. The responses were sorted according to nationality and published in 1908.

During this same period Henri Poincaré (1854–1912), one of the most noteworthy mathematicians of the time, had already laid much of the groundwork for his own pursuit of this same topic and in 1908 gave a presentation to the French Psychological Society in Paris entitled *L'Invention mathématique*—often mistranslated to *Mathematical Creativity*[1] (c.f. Poincaré 1952). At the time of the presentation Poincaré stated that he was aware of Claparède and Flournoy's work, as well as their results, but expressed that they would only confirm his own findings. Poincaré's presentation, as well as the essay it spawned, stands to this day as one of the most insightful, and thorough treatments of the topic of mathematical discovery, creativity, and invention.

> Just at this time, I left Caen, where I was living, to go on a geological excursion under the auspices of the School of Mines. The incident of the travel made me forget my mathematical work. Having reached Coutances, we entered an omnibus to go some place or other. At the moment when I put my foot on the step, the idea came to me, without anything in my former thoughts seeming to have paved the way for it, that the transformations I had used to define the Fuschian functions were identical with those of non-Euclidean geometry. I did not verify the idea; I should not have had the time, as, upon taking my seat in the omnibus, I went on with the conversation already commenced, but I felt a perfect certainty. On my return to Caen, for conscience' sake, I verified the results at my leisure. (Poincaré 1952, p. 53)

So powerful was his presentation, and so deep were his insights into his acts of invention and discovery that it could be said that he not so much described the characteristics of mathematical creativity, as defined them. From that point forth mathematical creativity, or even creativity in general, has not been discussed seriously without mention of Poincaré's name.

Inspired by this presentation, Jacques Hadamard (1865–1963), a contemporary and a friend of Poincaré's, began his own empirical investigation into this fascinating phenomenon. Hadamard had been critical of Claparède and Flournoy's work in that they had not adequately treated the topic on two fronts. As exhaustive as the survey appeared to be, Hadamard felt that it failed to ask some key questions—the most important of which was with regard to the reason for failures in the creation of

---

[1]Although it can be argued that there is a difference between creativity, discovery, and invention (see Liljedahl and Allan 2014) for the purposes of this book these will be assumed to be interchangeable.

mathematics. This seemingly innocuous oversight, however, led directly to his second and "most important criticism" (Hadamard 1945). He felt that only "first-rate men would dare to speak of" (p. 10) such failures. So, inspired by his good friend Poincaré's treatment of the subject Hadamard retooled the survey and gave it to friends of his for consideration—mathematicians such as Henri Poincaré and Albert Einstein, whose prominence were beyond reproach. Ironically, the new survey did not contain any questions that explicitly dealt with failure. In 1943 Hadamard gave a series of lectures on mathematical invention at the École Libre des Hautes Études in New York City. These talks were subsequently published as *The Psychology of Mathematical Invention in the Mathematical Field* (Hadameard 1945).

Hadamard's classic work treats the subject of invention at the crossroads of mathematics and psychology. It provides not only an entertaining look at the eccentric nature of mathematicians and their rituals, but also outlines the beliefs of mid twentieth-century mathematicians about the means by which they arrive at new mathematics. It is an extensive exploration and extended argument for the existence of unconscious mental processes. In essence, Hadamard took the ideas that Poincaré had posed and, borrowing a conceptual framework for the characterization of the creative process from the Gestaltists of the time (Wallas 1926), turned them into a stage theory. This theory still stands as the most viable and reasonable description of the process of mathematical creativity.

1.2.2 Defining Mathematical Creativity

The phenomena of mathematical creativity, although marked by sudden illumination, actually consist of four separate stages stretched out over time, of which illumination is but one stage. These stages are initiation, incubation, illumination, and verification (Hadamard 1945). The first of these stages, the *initiation* phase, consists of deliberate and conscious work. This would constitute a person's voluntary, and seemingly fruitless, engagement with a problem and be characterized by an attempt to solve the problem by trolling through a repertoire of past experiences. This is an important part of the inventive process because it creates the tension of unresolved effort that sets up the conditions necessary for the ensuing emotional release at the moment of illumination (Hadamard 1945; Poincaré 1952).

Following the initiation stage the solver, unable to come up with a solution stops working on the problem at a conscious level and begins to work on it at an unconscious level (Hadamard 1945; Poincaré 1952). This is referred to as the *incubation* stage of the inventive process and can last anywhere from several minutes to several years. After the period of incubation a rapid coming to mind of a solution, referred to as *illumination*, may occur. This is accompanied by a feeling of certainty and positive emotions (Poincaré 1952). Although the processes of incubation and illumination are shrouded behind the veil of the unconscious there are a number of things that can be deduced about them. First and foremost is the fact that unconscious work does, indeed, occur. Poincaré (1952), as well as Hadamard

(1945), use the very real experience of illumination, a phenomenon that cannot be denied, as evidence of unconscious work, the fruits of which appear in the flash of illumination. No other theory seems viable in explaining the sudden appearance of solution during a walk, a shower, a conversation, upon waking, or at the instance of turning the conscious mind back to the problem after a period of rest (Poincaré 1952). Also deducible is that unconscious work is inextricably linked to the conscious and intentional effort that precedes it.

> There is another remark to be made about the conditions of this unconscious work: it is possible, and of a certainty it is only fruitful, if it is on the one hand preceded and on the other hand followed by a period of conscious work. These sudden inspirations never happen except after some days of voluntary effort which has appeared absolutely fruitless and whence nothing good seems to have come ... (Poincaré 1952, p. 56)

Hence, the fruitless efforts of the initiation phase are only seemingly so. They not only set up the aforementioned tension responsible for the emotional release at the time of illumination, but also create the conditions necessary for the process to enter into the incubation phase.

Illumination is the manifestation of a bridging that occurs between the unconscious mind and the conscious mind (Poincaré 1952), a coming to (conscious) mind of an idea or solution. What brings the idea forward to consciousness is unclear, however. There are theories of the aesthetic qualities of the idea, effective surprise/shock of recognition, fluency of processing, or breaking functional fixedness. For reasons of brevity I will only expand on the first of these.

Poincaré proposed that ideas that were stimulated during initiation remained stimulated during incubation. However, freed from the constraints of conscious thought and deliberate calculation, these ideas would begin to come together in rapid and random unions so that "their mutual impacts may produce new combinations" (Poincaré 1952). These new combinations, or ideas, would then be evaluated for viability using an aesthetic sieve, which allows through to the conscious mind only the "right combinations" (Poincaré 1952). It is important to note, however, that good or aesthetic does not necessarily mean correct. Correctness is evaluated during the *verification* stage.

The purpose of verification is not only to check for correctness. It is also a method by which the solver re-engages with the problem at the level of details. That is, during the unconscious work the problem is engaged with at the level of ideas and concepts. During verification the solver can examine these ideas in closer details. Poincaré succinctly describes both of these purposes.

> As for the calculations, themselves, they must be made in the second period of conscious work, that which follows the inspiration, that in which one verifies the results of this inspiration and deduces their consequences. (Poincaré 1952, p. 62)

Aside from presenting this aforementioned theory on invention, Hadamard also engaged in a far-reaching discussion on a number of interesting, and sometimes quirky, aspects of invention and discovery that he had culled from the results of his empirical study, as well as from pertinent literature. This discussion was nicely summarized by Newman (2000) in his commentary on the elusiveness of invention.

> The celebrated phrenologist Gall said mathematical ability showed itself in a bump on the head, the location of which he specified. The psychologist Souriau, we are told, maintained that invention occurs by "pure chance", a valuable theory. It is often suggested that creative ideas are conjured up in "mathematical dreams", but this attractive hypothesis has not been verified. Hadamard reports that mathematicians were asked whether "noises" or "meteorological circumstances" helped or hindered research [..] Claude Bernard, the great physiologist, said that in order to invent "one must think aside". Hadamard says this is a profound insight; he also considers whether scientific invention may perhaps be improved by standing or sitting or by taking two baths in a row. Helmholtz and Poincaré worked sitting at a table; Hadamard's practice is to pace the room ("Legs are the wheels of thought", said Emile Angier); the chemist J. Teeple was the two-bath man. (p. 2039)

1.2.3 Discourses on Creativity

Creativity is a term that can be used both loosely and precisely. That is, while there exists a common usage of the term there also exists a tradition of academic discourse on the subject. A common usage of *creative* refers to a process or a person whose products are original, novel, unusual, or even abnormal (Csíkszentmihályi 1996). In such a usage, creativity is assessed on the basis of the external and observable products of the process, the process by which the product comes to be, or on the character traits of the person doing the 'creating'. Each of these usages—product, process, person—is the roots of the discourses (Liljedahl and Allan 2014) that I summarize here, the first of which concerns products.

Consider a mother who states that her daughter is creative because she drew an original picture. The basis of such a statement can lie either in the fact that the picture is unlike any the mother has ever seen or unlike any her daughter has ever drawn before. This mother is assessing creativity on the basis of what her daughter has produced. However, the standards that form the basis of her assessment are neither consistent nor stringent. There does not exist a universal agreement as to what she is comparing the picture to (pictures by other children or other pictures by the same child). Likewise, there is no standard by which the actual quality of the picture is measured. The academic discourse that concerns assessment of products, on the other hand, is both consistent and stringent (Csíkszentmihályi 1996). This discourse concerns itself more with a fifth, and as yet unmentioned, stage of the creative process; *elaboration*. Elaboration is where inspiration becomes perspiration (Csíkszentmihályi 1996). It is the act of turning a good idea into a finished product, and the finished product is ultimately what determines the *creativity* of the process that spawned it—that is, it cannot be a creative process if nothing is created. In particular, this discourse demands that the product be assessed against other products within its field, by the members of that field, to determine if it is original AND useful (Csíkszentmihályi 1996; Bailin 1994). If it is, then the product is deemed to be creative. Note that such a use of assessment of end product pays very little attention to the actual process that brings this product forth.

The second discourse concerns the creative process. The literature pertaining to this can be separated into two categories—a prescriptive discussion of the creativity process and a descriptive discussion of the creativity process. Although both of these discussions have their roots in the four stages that Wallas (1926) proposed makes up the creative process, they make use of these stages in very different ways. The prescriptive discussion of the creative process is primarily focused on the first of the four stages, *initiation*, and is best summarized as a *cause-and-effect* discussion of creativity, where the thinking processes during the initiation stage are the *cause* and the creative outcome are the *effects* (Ghiselin 1952). Some of the literature claims that the seeds of creativity lie in being able to think about a problem or situation analogically. Other literature claims that utilizing specific thinking tools such as imagination, empathy, and embodiment will lead to creative products. In all of these cases, the underlying theory is that the eventual presentation of a creative idea will be precipitated by the conscious and deliberate efforts during the initiation stage. On the other hand, the literature pertaining to a descriptive discussion of the creative process is inclusive of all four stages (Kneller 1965; Koestler 1964). For example, Csíkszentmihályi (1996), in his work on *flow* attends to each of the stages, with much attention paid to the fluid area between conscious and unconscious work, or initiation and incubation. His claim is that the creative process is intimately connected to the enjoyment that exists during times of sincere and consuming engagement with a situation, the conditions of which he describes in great detail.

The third, and final, discourse on creativity pertains to the person. This discourse is dominated by two distinct characteristics, habit and genius. Habit has to do with the personal habits as well as the habits of mind of people that have been deemed to be creative. However, creative people are most easily identified through their reputation for genius. Consequently, this discourse is often dominated by the analyses of the habits of geniuses as is seen in the work of Ghiselin (1952), Koestler (1964), and Kneller (1965) who draw on historical personalities such as Albert Einstein, Henri Poincaré, Vincent Van Gogh, D.H. Lawrence, Samuel Taylor Coleridge, Igor Stravinsky, and Wolfgang Amadeus Mozart to name a few. The result of this sort of treatment is that creative acts are viewed as rare mental feats, which are produced by extraordinary individuals who use extraordinary thought processes.

These different discourses on creativity can be summed up in a tension between absolutist and relativist perspectives on creativity (Liljedahl and Sriraman 2006). An absolutist perspective assumes that creative processes are the domain of genius and are present only as precursors to the creation of remarkably useful and universally novel products. The relativist perspective, on the other hand, allows for every individual to have moments of creativity that may, or may not, result in the creation of a product that may, or may not, be either useful or novel.

> Between the work of a student who tries to solve a problem in geometry or algebra and a work of invention, one can say there is only a difference of degree. (Hadamard 1945, p. 104).

Regardless of discourse, however, creativity is not "part of the theories of logical forms" (Dewey 1938). That is, creativity is not representative of the lock-step logic and deductive reasoning that mathematical problem solving is often presumed to embody (Bibby 2002; Burton 1999). Couple this with the aforementioned demanding constraints as to what constitutes a problem, where then does that leave problem solving heuristics? More specifically, are there *creative* problem solving heuristics that will allow us to resolve problems that require illumination to solve? The short answer to this question is yes—there does exist such problem solving heuristics. To understand these, however, we must first understand the routine problem solving heuristics they are built upon. In what follows, I walk through the work of key authors and researchers whose work offers us insights into progressively more creative problem solving heuristics for solving true problems.

1.2.4 Problem Solving by Design

In a general sense, design is defined as the algorithmic and deductive approach to solving a problem (Rusbult 2000). This process begins with a clearly defined goal or objective after which there is a great reliance on relevant past experience, referred to as repertoire (Bruner 1964; Schön 1987), to produce possible options that will lead towards a solution of the problem (Poincaré 1952). These options are then examined through a process of conscious evaluations (Dewey 1933) to determine their suitability for advancing the problem towards the final goal. In very simple terms, problem solving by design is the process of deducing the solution from that which is already known.

Mayer (1982), Schoenfeld (1982), and Silver (1982) state that prior knowledge is a key element in the problem solving process. Prior knowledge influences the problem solver's understanding of the problem as well as the choice of strategies that will be called upon in trying to solve the problem. In fact, prior knowledge and prior experiences is all that a solver has to draw on when first attacking a problem. As a result, all problem solving heuristics incorporate this resource of past experiences and prior knowledge into their initial attack on a problem. Some heuristics refine these ideas, and some heuristics extend them (c.f. Kilpatrick 1985; Bruder 2000). Of the heuristics that refine, none is more influential than the one created by George Pólya (1887–1985).

1.2.5 George Pólya: How to Solve It

In his book *How to Solve It* (1949) Pólya lays out a problem solving heuristic that relies heavily on a repertoire of past experience. He summarizes the four-step process of his heuristic as follows:

1. Understanding the Problem

- First. You have to understand the problem.
- What is the unknown? What are the data? What is the condition?
- Is it possible to satisfy the condition? Is the condition sufficient to determine the unknown? Or is it insufficient? Or redundant? Or contradictory?
- Draw a figure. Introduce suitable notation.
- Separate the various parts of the condition. Can you write them down?

2. Devising a Plan

- Second. Find the connection between the data and the unknown. You may be obliged to consider auxiliary problems if an immediate connection cannot be found. You should obtain eventually a plan of the solution.
- Have you seen it before? Or have you seen the same problem in a slightly different form?
- Do you know a related problem? Do you know a theorem that could be useful?
- Look at the unknown! And try to think of a familiar problem having the same or a similar unknown.
- Here is a problem related to yours and solved before. Could you use it? Could you use its result? Could you use its method? Should you introduce some auxiliary element in order to make its use possible?
- Could you restate the problem? Could you restate it still differently? Go back to definitions.
- If you cannot solve the proposed problem try to solve first some related problem. Could you imagine a more accessible related problem? A more general problem? A more special problem? An analogous problem? Could you solve a part of the problem? Keep only a part of the condition, drop the other part; how far is the unknown then determined, how can it vary? Could you derive something useful from the data? Could you think of other data appropriate to determine the unknown? Could you change the unknown or data, or both if necessary, so that the new unknown and the new data are nearer to each other?
- Did you use all the data? Did you use the whole condition? Have you taken into account all essential notions involved in the problem?

3. Carrying Out the Plan

- Third. Carry out your plan.
- Carrying out your plan of the solution, check each step. Can you see clearly that the step is correct? Can you prove that it is correct?

4. Looking Back

- Fourth. Examine the solution obtained.
- Can you check the result? Can you check the argument?

- Can you derive the solution differently? Can you see it at a glance?
- Can you use the result, or the method, for some other problem?

The emphasis on auxiliary problems, related problems, and analogous problems that are, in themselves, also familiar problems is an explicit manifestation of relying on a repertoire of past experience. This use of familiar problems also requires an ability to deduce from these related problems a recognizable and relevant attribute that will transfer to the problem at hand. The mechanism that allows for this transfer of knowledge between analogous problems is known as analogical reasoning (English 1997, 1998; Novick 1988, 1990, 1995; Novick and Holyoak 1991) and has been shown to be an effective, but not always accessible, thinking strategy.

Step four in Pólya's heuristic, looking back, is also a manifestation of utilizing prior knowledge to solve problems, albeit an implicit one. Looking back makes connections "in memory to previously acquired knowledge [..] and further establishes knowledge in long-term memory that may be elaborated in later problem-solving encounters" (Silver 1982, p. 20). That is, looking back is a forward-looking investment into future problem solving encounters, it sets up connections that may later be needed.

Pólya's heuristic is a refinement on the principles of problem solving by design. It not only makes explicit the focus on past experiences and prior knowledge, but also presents these ideas in a very succinct, digestible, and teachable manner. This heuristic has become a popular, if not the most popular, mechanism by which problem solving is taught and learned.

1.2.6 Alan Schoenfeld: Mathematical Problem Solving

The work of Alan Schoenfeld is also a refinement on the principles of problem solving by design. However, unlike Pólya (1949) who refined these principles at a theoretical level, Schoenfeld has refined them at a practical and empirical level. In addition to studying taught problem solving strategies he has also managed to identify and classify a variety of strategies, mostly ineffectual, that students invoke naturally (Schoenfeld 1985, 1992). In so doing, he has created a better understanding of how students solve problems, as well as a better understanding of how problems should be solved and how problem solving should be taught.

For Schoenfeld, the problem solving process is ultimately a dialogue between the problem solver's prior knowledge, his attempts, and his thoughts along the way (Schoenfeld 1982). As such, the solution path of a problem is an emerging and contextually dependent process. This is a departure from the predefined and contextually independent processes of Pólya's (1949) heuristics. This can be seen in Schoenfeld's (1982) description of a good problem solver.

> To examine what accounts for expertise in problem solving, you would have to give the expert a problem for which he does not have access to a solution schema. His behavior in such circumstances is radically different from what you would see when he works on routine or familiar "non-routine" problems. On the surface his performance is no longer

proficient; it may even seem clumsy. Without access to a solution schema, he has no clear indication of how to start. He may not fully understand the problem, and may simply "explore it for a while until he feels comfortable with it. He will probably try to "match" it to familiar problems, in the hope it can be transformed into a (nearly) schema-driven solution. He will bring up a variety of plausible things: related facts, related problems, tentative approaches, etc. All of these will have to be juggled and balanced. He may make an attempt solving it in a particular way, and then back off. He may try two or three things for a couple of minutes and then decide which to pursue. In the midst of pursuing one direction he may go back and say "that's harder than it should be" and try something else. Or, after the comment, he may continue in the same direction. With luck, after some aborted attempts, he will solve the problem. (p. 32-33)

Aside from demonstrating the emergent nature of the problem solving process, this passage also brings forth two consequences of Schoenfeld's work. The first of these is the existence of problems for which the solver does not have "access to a solution schema". Unlike Pólya (1949), who's heuristic is a 'one size fits all (problems)' heuristic, Schoenfeld acknowledges that problem solving heuristics are, in fact, personal entities that are dependent on the solver's prior knowledge as well as their understanding of the problem at hand. Hence, the problems that a person can solve through his or her personal heuristic are finite and limited.

The second consequence that emerges from the above passage is that if a person lacks the solution schema to solve a given problem s/he may still solve the problem with the help of *luck*. This is an acknowledgement, if only indirectly so, of the difference between problem solving in an intentional and mechanical fashion verses problem solving in a more creative fashion, which is neither intentional nor mechanical (Pehkonen 1997).

1.2.7 David Perkins: Breakthrough Thinking

As mentioned, many consider a problem that can be solved by intentional and mechanical means to not be worthy of the title 'problem'. As such, a repertoire of past experiences sufficient for dealing with such a 'problem' would disqualify it from the ranks of 'problems' and relegate it to that of 'exercises'. For a problem to be classified as a 'problem', then, it must be 'problematic'. Although such an argument is circular it is also effective in expressing the ontology of mathematical 'problems'.

Perkins (2000) also requires problems to be problematic. His book *Archimedes' Bathtub: The Art and Logic of Breakthrough Thinking* (2000) deals with situations in which the solver has gotten stuck and no amount of intentional or mechanical adherence to the principles of past experience and prior knowledge is going to get them unstuck. That is, he deals with problems that, by definition, cannot be solved through a process of design [or through the heuristics proposed by Pólya (1949) and Schoenfeld (1985)]. Instead, the solver must rely on the extra-logical process of what Perkins (2000) calls *breakthrough thinking*.

Perkins (2000) begins by distinguishing between reasonable and unreasonable problems. Although both are solvable, only reasonable problems are solvable through reasoning. Unreasonable problems require a breakthrough in order to solve them. The problem, however, is itself inert. It is neither reasonable nor unreasonable. That quality is brought to the problem by the solver. That is, if a student cannot solve a problem by direct effort then that problem is deemed to be unreasonable for that student. Perkins (2000) also acknowledges that what is an unreasonable problem for one person is a perfectly reasonable problem for another person; reasonableness is dependent on the person.

This is not to say that, once found, the solution cannot be seen as accessible through reason. During the actual process of solving, however, direct and deductive reasoning does not work. Perkins (2000) uses several classic examples to demonstrate this, the most famous being the problem of connecting nine dots in a 3 × 3 array with four straight lines without removing pencil from paper, the solution to which is presented in Fig. 1.

To solve this problem, Perkins (2000) claims that the solver must recognize that the constraint of staying within the square created by the 3 × 3 array is a self-imposed constraint. He further claims that until this is recognized no amount of reasoning is going to solve the problem. That is, at this point in the problem solving process the problem is unreasonable. However, once this self-imposed constraint is recognized the problem, and the solution, are perfectly reasonable. Thus, the solution of an, initially, unreasonable problem is reasonable.

The problem solving heuristic that Perkins (2000) has constructed to deal with solvable, but unreasonable, problems revolves around the idea of breakthrough thinking and what he calls *breakthrough problems*. A breakthrough problem is a solvable problem in which the solver has gotten stuck and will require an AHA! to get unstuck and solve the problem. Perkins (2000) poses that there are only four types of solvable unreasonable problems, which he has named *wilderness of possibilities*, the *clueless plateau*, *narrow canyon of exploration*, and *oasis of false promise*. The names for the first three of these types of problems are related to the Klondike gold rush in Alaska, a time and place in which gold was found more by luck than by direct and systematic searching.

The wilderness of possibilities is a term given to a problem that has many tempting directions but few actual solutions. This is akin to a prospector searching

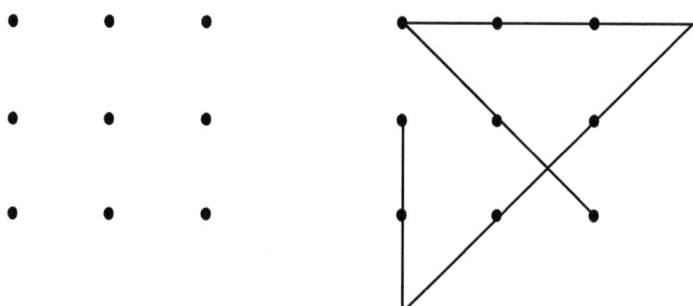

Fig. 1 Nine dots—four lines problem and solution

for gold in the Klondike. There is a great wilderness in which to search, but very little gold to be found. The clueless plateau is given to problems that present the solver with few, if any, clues as to how to solve it. The narrow canyon of exploration is used to describe a problem that has become constrained in such a way that no solution now exists. The nine-dot problem presented above is such a problem. The imposed constraint that the lines must lie within the square created by the array makes a solution impossible. This is identical to the metaphor of a prospector searching for gold within a canyon where no gold exists. The final type of problem gets its name from the desert. An oasis of false promise is a problem that allows the solver to quickly get a solution that is close to the desired outcome; thereby tempting them to remain fixed on the strategy that they used to get this almost-answer. The problem is, that like the canyon, the solution does not exist at the oasis; the solution strategy that produced an almost-answer is incapable of producing a complete answer. Likewise, a desert oasis is a false promise in that it is only a reprieve from the desolation of the dessert and not a final destination.

Believing that there are only four ways to get stuck, Perkins (2000) has designed a problem solving heuristic that will "up the chances" of getting unstuck. This heuristic is based on what he refers to as "the logic of lucking out" (p. 44) and is built on the idea of introspection. By first recognizing that they are stuck, and then recognizing that the reason they are stuck can only be attributed to one of four reasons, the solver can access four strategies for getting unstuck, one each for the type of problem they are dealing with. If the reason they are stuck is because they are faced with a wilderness of possibilities they are to begin roaming far, wide, and systematically in the hope of reducing the possible solution space to one that is more manageable. If they find themselves on a clueless plateau they are to begin looking for clues, often in the wording of the problem. When stuck in a narrow canyon of possibilities they need to re-examine the problem and see if they have imposed any constraints. Finally, when in an oasis of false promise they need to re-attack the problem in such a way that they stay away from the oasis.

Of course, there are nuances and details associated with each of these types of problems and the strategies for dealing with them. However, nowhere within these details is there mention of the main difficulty inherent in introspection; that it is much easier for the solver to get stuck than it is for them to recognize that they are stuck. Once recognized, however, the details of Perkins' (2000) heuristic offer the solver some ways for recognizing why they are stuck.

1.2.8 John Mason, Leone Burton, and Kaye Stacey: Thinking Mathematically

The work of Mason et al. in their book *Thinking Mathematically* (1982) also recognizes the fact that for each individual there exists problems that will not yield to their intentional and mechanical attack. The heuristic that they present for dealing with this has two main processes with a number of smaller phases, rubrics, and states. The main processes are what they refer to as specializing and generalizing.

Specializing is the process of getting to know the problem and how it behaves through the examination of special instances of the problem. This process is synonymous with problem solving by design and involves the repeated oscillation between the entry and attack phases of Mason et al. (1982) heuristic. The entry phase is comprised of 'getting started' and 'getting involved' with the problem by using what is immediately known about it. Attacking the problem involves conjecturing and testing a number of hypotheses in an attempt to gain greater understanding of the problem and to move towards a solution.

At some point within this process of oscillating between entry and attack the solver will get stuck, which Mason et al. (1982) refer to as "an honourable and positive state, from which much can be learned" (p. 55). The authors dedicate an entire chapter to this state in which they acknowledge that getting stuck occurs long before an awareness of being stuck develops. They proposes that the first step to dealing with being stuck is the simple act of writing STUCK!

> The act of expressing my feelings helps to distance me from my state of being stuck. It frees me from incapacitating emotions and reminds me of actions that I can take. (p. 56)

The next step is to reengage the problem by examining the details of what is known, what is wanted, what can be introduced into the problem, and what has been introduced into the problem (imposed assumptions). This process is engaged in until an AHA!, which advances the problem towards a solution, is encountered. If, at this point, the problem is not completely solved the oscillation is then resumed.

At some point in this process an attack on the problem will yield a solution and generalizing can begin. Generalizing is the process by which the specifics of a solution are examined and questions as to why it worked are investigated. This process is synonymous with the verification and elaboration stages of invention and creativity. Generalization may also include a phase of review that is similar to Pólya's (1949) looking back.

1.2.9 Gestalt: The Psychology of Problem Solving

The Gestalt psychology of learning believes that all learning is based on insights (Koestler 1964). This psychology emerged as a response to behaviourism, which claimed that all learning was a response to external stimuli. Gestalt psychologists, on the other hand, believed that there was a cognitive process involved in learning as well. With regards to problem solving, the Gestalt school stands firm on the belief that problem solving, like learning, is a product of insight and as such, cannot be taught. In fact, the theory is that not only can problem solving not be taught, but also that attempting to adhere to any sort of heuristic will impede the working out of a correct solution (Krutestkii 1976). Thus, there exists no Gestalt problem solving heuristic. Instead, the practice is to focus on the problem and the solution rather than on the process of coming up with a solution. Problems are solved by turning them over and over in the mind until an insight, a viable avenue of attack, presents itself. At the same time, however, there is a great reliance on prior knowledge and past experiences. The Gestalt method of problem solving, then, is at the same time very different and very similar to the process of design.

Gestalt psychology has not fared well during the evolution of cognitive psychology. Although it honours the work of the unconscious mind it does so at the expense of practicality. If learning is, indeed, entirely based on insight then there is little point in continuing to study learning. "When one begins by assuming that the most important cognitive phenomena are inaccessible, there really is not much left to talk about" (Schoenfeld 1985, p. 273). However, of interest here is the Gestalt psychologists' claim that focus on problem solving methods creates functional fixedness (Ashcraft 1989). Mason et al. (1982), as well as Perkins (2000) deal with this in their work on getting unstuck.

1.2.10   Final Comments

Mathematics has often been characterized as the most precise of all sciences. Lost in such a misconception is the fact that mathematics often has its roots in the fires of creativity, being born of the extra-logical processes of illumination and intuition. Problem solving heuristics that are based solely on the processes of logical and deductive reasoning distort the true nature of problem solving. Certainly, there are problems in which logical deductive reasoning is sufficient for finding a solution. But these are not true problems. True problems need the extra-logical processes of creativity, insight, and illumination, in order to produce solutions.

Fortunately, as elusive as such processes are, there does exist problem solving heuristics that incorporate them into their strategies. Heuristics such as those by Perkins (2000) and Mason et al. (1982) have found a way of combining the intentional and mechanical processes of problem solving by design with the extra-logical processes of creativity, illumination, and the AHA!. Furthermore, they have managed to do so without having to fully comprehend the inner workings of this mysterious process.

1.3   Digital Technologies and Mathematical Problem Solving—Luz Manuel Santos-Trigo

Mathematical problem solving is a field of research that focuses on analysing the extent to which problem solving activities play a crucial role in learners' understanding and use of mathematical knowledge. Mathematical problems are central in mathematical practice to develop the discipline and to foster students learning (Pólya 1945; Halmos 1994). Mason and Johnston-Wilder (2006) pointed out that "The purpose of a task is to initiate mathematically fruitful activity that leads to a transformation in what learners are sensitized to notice and competent to carry out"

(p. 25). Tasks are essential for learners to elicit their ideas and to engage them in mathematical thinking. In a problem solving approach, what matters is the learners' goals and ways to interact with the tasks. That is, even routine tasks can be a departure point for learners to extend initial conditions and transform them into some challenging activities.

Thus, analysing and characterizing ways in which mathematical problems are formulated (Singer et al. 2015) and the process involved in pursuing and solving those problems generate important information to frame and structure learning environments to guide and foster learners' construction of mathematical concepts and problem solving competences (Santos-Trigo 2014). Furthermore, mathematicians or discipline practitioners have often been interested in unveiling and sharing their own experience while developing the discipline. As a results, they have provided valuable information to characterize mathematical practices and their relations to what learning processes of the discipline entails. It is recognized that the work of Pólya (1945) offered not only bases to launch several research programs in problem solving (Schoenfeld 1992; Mason et al. 1982); but also it became an essential resource for teachers to orient and structure their mathematical lessons (Krulik and Reys 1980).

1.3.1 Research Agenda

A salient feature of a problem solving approach to learn mathematics is that teachers and students develop and apply an enquiry or inquisitive method to delve into mathematical concepts and tasks. How are mathematical problems or concepts formulated? What types of problems are important for teachers/learners to discuss and engage in mathematical reasoning? What mathematical processes and ways of reasoning are involved in understanding mathematical concepts and solving problems? What are the features that distinguish an instructional environment that fosters problem-solving activities? How can learners' problem solving competencies be assessed? How can learners' problem solving competencies be characterized and explained? How can learners use digital technologies to understand mathematics and to develop problem-solving competencies? What ways of reasoning do learners construct when they use digital technologies in problem solving approaches? These types of questions have been important in the problem solving research agenda and delving into them has led researchers to generate information and results to support and frame curriculum proposals and learning scenarios. The purpose of this section is to present and discuss important themes that emerged in problem solving approaches that rely on the systematic use of several digital technologies.

In the last 40 years, the accumulated knowledge in the problem solving field has shed lights on both a characterization of what mathematical thinking involves and how learners can construct a robust knowledge in problem solving environments (Schoenfeld 1992). In this process, the field has contributed to identify what types of transformations traditional learning scenarios might consider when teachers and students incorporate the use of digital technologies in mathematical classrooms. In this context, it is important to briefly review what main themes and developments the field has addressed and achieved during the last 40 years.

1.3.2 Problem Solving Developments

There are traces of mathematical problems and solutions throughout the history of civilization that explain the humankind interest for identifying and exploring mathematical relations (Kline 1972). Pólya (1945) reflects on his own practice as a mathematician to characterize the process of solving mathematical problems through four main phases: Understanding the problem, devising a plan, carrying out the plan, and looking back. Likewise, Pólya (1945) presents and discusses the role played by heuristic methods throughout all problem solving phases. Schoenfeld (1985) presents a problem solving research program based on Pólya's (1945) ideas to investigate the extent to which problem solving heuristics help university students to solve mathematical problems and to develop a way of thinking that shows consistently features of mathematical practices. As a result, he explains the learners' success or failure in problem solving activities can be characterized in terms their mathematical resources and ways to access them, cognitive and metacognitive strategies used to represent and explore mathematical tasks, and systems of beliefs about mathematics and solving problems. In addition, Schoenfeld (1992) documented that heuristics methods as illustrated in Pólya's (1945) book are ample and general and do not include clear information and directions about how learners could assimilate, learn, and use them in their problem solving experiences. He suggested that students need to discuss what it means, for example, to think of and examining special cases (one important heuristic) in finding a closed formula for series or sequences, analysing relationships of roots of polynomials, or focusing on regular polygons or equilateral/right triangles to find general relations about these figures. That is, learners need to work on examples that lead them to recognize that the use of a particular heuristic often involves thinking of different type of cases depending on the domain or content involved. Lester and Kehle (2003) summarize themes and methodological shifts in problem solving research up to 1995. Themes include what makes a problem difficult for students and what it means to be successful problem solvers; studying and contrasting experts and novices' problem solving approaches; learners' metacognitive, beliefs systems and the influence of affective behaviours; and the role of context; and social interactions in problem solving environments. Research methods in problem solving studies have gone from emphasizing quantitative or statistical design to the use of cases studies and ethnographic methods (Krutestkii (1976). Teaching strategies also evolved from being centred on teachers to the active students' engagement and collaboration approaches (NCTM 2000). Lesh and Zawojewski (2007) propose to extend problem solving approaches beyond class setting and they introduce the construct "model eliciting activities" to delve into the learners' ideas and thinking as a way to engage them in the development of problem solving experiences. To this end, learners develop and constantly refine problem-solving competencies as a part of a learning community that promotes and values modelling construction activities. Recently, English and Gainsburg (2016) have discussed the importance of modeling eliciting activities to prepare and develop students' problem solving experiences for 21st Century challenges and demands.

Törner et al. (2007) invited mathematics educators worldwide to elaborate on the influence and developments of problem solving in their countries. Their contributions show a close relationship between countries mathematical education traditions and ways to frame and implement problem solving approaches. In Chinese classrooms, for example, three instructional strategies are used to structure problem solving lessons: *one problem multiple solutions*, *multiple problems one solution*, and *one problem multiple changes*. In the Netherlands, the realistic mathematical approach permeates the students' development of problem solving competencies; while in France, problem solving activities are structured in terms of two influential frameworks: The theory of didactical situations and anthropological theory of didactics.

In general, problem solving frameworks and instructional approaches came from analysing students' problem solving experiences that involve or rely mainly on the use of paper and pencil work. Thus, there is a need to re-examined principles and frameworks to explain what learners develop in learning environments that incorporate systematically the coordinated use of digital technologies (Hoyles and Lagrange 2010). In this perspective, it becomes important to briefly describe and identify what both multiple purpose and ad hoc technologies can offer to the students in terms of extending learning environments and representing and exploring mathematical tasks. Specifically, a task is used to identify features of mathematical reasoning that emerge through the use digital technologies that include both mathematical action and multiple purpose types of technologies.

1.3.3 Background

Digital technologies are omnipresent and their use permeates and shapes several social and academic events. Mobile devices such as tablets or smart phones are transforming the way people communicate, interact and carry out daily activities. Churchill et al. (2016) pointed out that mobile technologies provide a set of tools and affordances to structure and support learning environments in which learners continuously interact to construct knowledge and solve problems. The tools include resources or online materials, efficient connectivity to collaborate and discuss problems, ways to represent, explore and store information, and analytical and administration tools to management learning activities. Schmidt and Cohen (2013) stated that nowadays it is difficult to imagine a life without mobile devices, and communication technologies are playing a crucial role in generating both cultural and technical breakthroughs. In education, the use of mobile artefacts and computers offers learners the possibility of continuing and extending peers and groups' mathematical discussions beyond formal settings. In this process, learners can also consult online materials and interact with experts, peers or more experienced students while working on mathematical tasks. In addition, dynamic geometry systems (GeoGebra) provide learners a set of affordances to represent and explore dynamically mathematical problems. Leung and Bolite-Frant (2015) pointed out that tools help activate an interactive environment in which teachers and students' mathematical experiences get enriched. Thus, the digital age brings new challenges to the mathematics education community related to the changes that technologies produce to curriculum, learning scenarios, and ways to represent, explore mathematical situations. In particular, it is important to characterize the type of reasoning that learners can develop as a result of using digital technologies in their process of learning concepts and solving mathematical problems.

1.3.4   A Focus on Mathematical Tasks

Mathematical tasks are essential elements for engaging learners in mathematical reasoning which involves representing objects, identifying and exploring their properties in order to detect invariants or relationships and ways to support them. Watson and Ohtani (2015) stated that task design involves discussions about mathematical content and students' learning (cognitive perspective), about the students' experiences to understand the nature of mathematical activities; and about the role that tasks played in teaching practices. In this context, tasks are the vehicle to present and discuss theoretical frameworks for supporting the use of digital technology, to analyse the importance of using digital technologies in extending learners' mathematical discussions beyond formal settings, and to design ways to foster and assess the use of technologies in learners' problem solving environments. In addition, it is important to discuss contents, concepts, representations and strategies involved in the process of using digital technologies in approaching the tasks. Similarly, it becomes essential to discuss what types of activities students will do to learn and solve the problems in an environment where the use of technologies fosters and values the participation and collaboration of all students. What digital technologies are important to incorporate in problem solving approaches? Dynamic Geometry Systems can be considered as a milestone in the development of digital technologies. Objects or mathematical situations can be represented dynamically through the use of a Dynamic Geometry System and learners or problem solvers can identify and examine mathematical relations that emerge from moving objects within the dynamic model (Moreno-Armella and Santos-Trigo 2016).

Leung and Bolite-Frant (2015) stated that "dynamic geometry software can be used in task design to cover a large epistemic spectrum from drawing precise robust geometrical figures to exploration of new geometric theorems and development of argumentation discourse" (p. 195). As a result, learners not only need to develop skills and strategies to construct dynamic configuration of problems; but also ways of relying on the tool's affordances (quantifying parameters or objects attributes, generating loci, graphing objects behaviours, using sliders, or dragging particular elements within the configuration) in order to identify and support mathematical relations. What does it mean to represent and explore an object or mathematical situation dynamically?

A simple task that involves a rhombus and its inscribed circle is used to illustrate how a dynamic representation of these objects and embedded elements can lead learners to identify and examine mathematical properties of those objects in the construction of the configuration. To this end, learners are encouraged to pose and pursue questions to explain the behaviours of parameters or attributes of the family of objects that is generated as a result of moving a particular element within the configuration.

1.3.5 A Task: A Dynamic Rhombus

Figure 2 represents a rhombus APDB and its inscribed circle (O is intersection of diagonals AD and BP and the radius of the inscribed circle is the perpendicular segment from any side of the rhombus to point O), vertex P lies on a circle c centred at point A. Circle c is only a heuristic to generate a family of rhombuses. Thus, point P can be moved along circle c to generate a family of rhombuses. Indeed, based on the symmetry of the circle it is sufficient to move P on the semicircle B'CA to draw such a family of rhombuses.

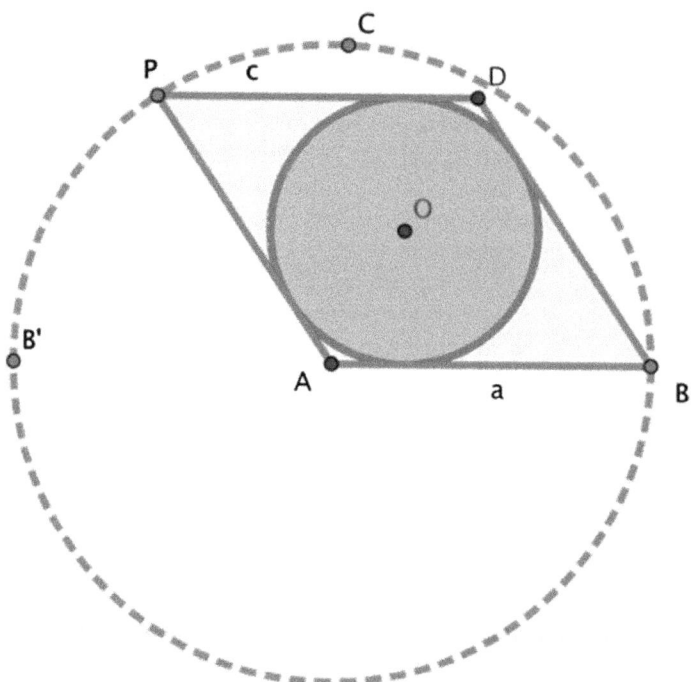

Fig. 2 A dynamic construction of a rhombus

1.3.6   Posing Questions

A goal in constructing a dynamic model or configuration of problems is always to identify and explore mathematical properties and relations that might result from moving objects within the model. How do the areas of both the rhombus and the inscribed circle behave when point P is moved along the arc B'CB? At what position of point P does the area of the rhombus or inscribed circle reach the maximum value? The coordinates of points S and Q (Fig. 3) are the *x*-value of point

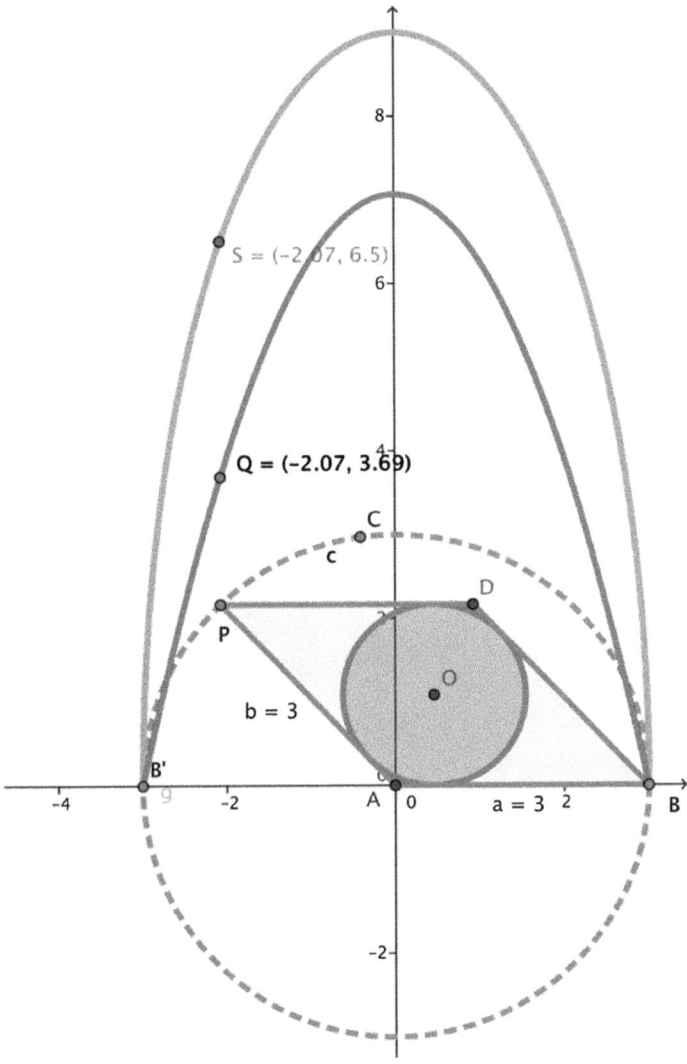

Fig. 3 Graphic representation of the area variation of the family of rhombuses and inscribed circles generated when P is moved through arc B'CB

P and as y-value the corresponding area values of rhombus ABDP and the inscribed circle respectively. Figure 2 shows the loci of points S and Q when point P is moved along arc B'CB. Here, finding the locus via the use of GeoGebra is another heuristic to graph the area behaviour without making explicit the algebraic model of the area.

The area graphs provide information to visualize that in that family of generated rhombuses the maximum area value of the inscribed circle and rhombus is reached when the rhombus becomes a square (Fig. 4). That is, the controlled movement of

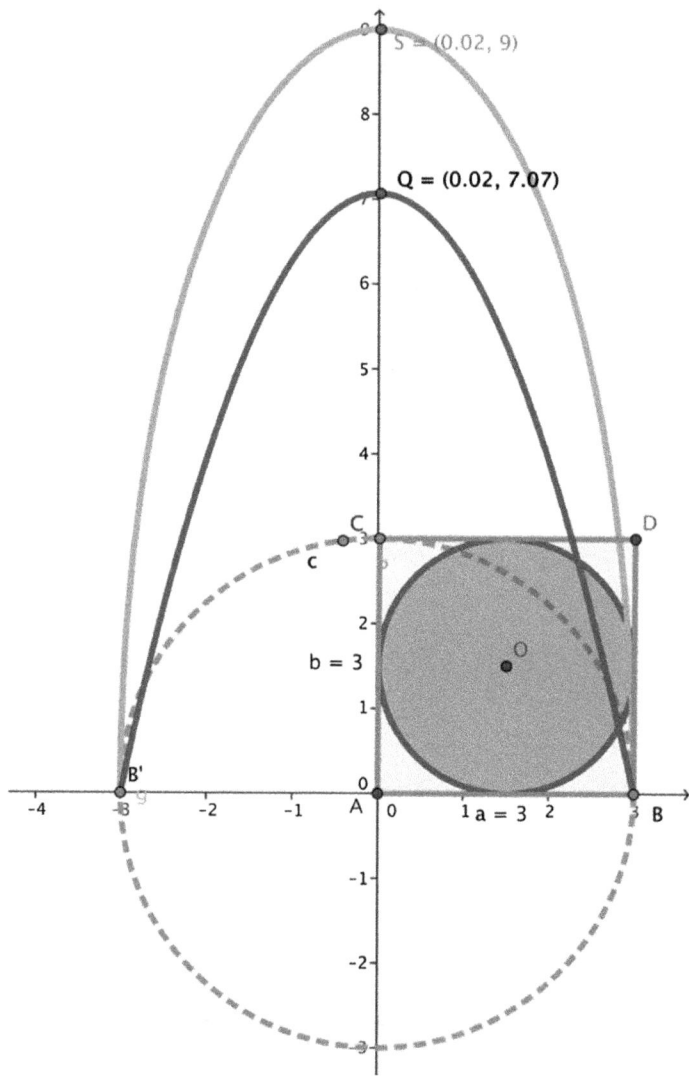

Fig. 4 Visualizing the rhombus and the inscribed circle with maximum area

particular objects is an important strategy to analyse the area variation of the family of rhombuses and their inscribed circles.

It is important to observe the identification of points P and Q in terms of the position of point P and the corresponding areas and the movement of point P was sufficient to generate both area loci. That is, the graph representation of the areas is achieved without having an explicit algebraic expression of the area variation. Clearly, the graphic representations provide information regarding the increasing or decreasing interval of both areas; it is also important to explore what properties both graphic representations hold. The goal is to argue that the area variation of the rhombus represents an ellipse and the area of the inscribed circle represents a parabola. An initial argument might involve selecting five points on each locus and using the tool to draw the corresponding conic section (Fig. 5). In this case, the tool affordances play an important role in generating the graphic representation of the areas' behaviours and in identifying properties of those representations. In this context, the use of the tool can offer learners the opportunity to problematize (Santos-Trigo 2007) a simple mathematical object (rhombus) as a means to search for mathematical relations and ways to support them.

1.3.7 Looking for Different Solutions Methods

Another line of exploration might involve asking for ways to construct a rhombus and its inscribed circle: Suppose that the side of the rhombus and the circle are given, how can you construct the rhombus that has that circle inscribed? Figure 6 shows the given data, segment A_1B_1 and circle centred at O and radius OD. The initial goal is to draw the circle tangent to the given segment. To this end, segment AB is congruent to segment A_1B_1 and on this segment a point P is chosen and a perpendicular to segment AB that passes through point P is drawn. Point C is on this perpendicular and the centre of a circle with radius OD and h is the perpendicular to line PC that passes through point C. Angle ACB changes when point P is moved along segment AB and point E and F are the intersection of line h and the circle with centre M the midpoint of AB and radius MA (Fig. 6).

Figure 7a shows the right triangle AFB as the base to construct the rhombus and the inscribed circle and Fig. 7b shows the second solution based on triangle AEB.

Another approach might involve drawing the given circle centred at the origin and the segment as EF with point E on the y-axis. Line OC is perpendicular to segment EF and the locus of point C when point E moves along the y-axis intersects the given circle (Fig. 8a, b). Both figures show two solutions to draw the rhombus that circumscribe the given circle.

In this example, the GeoGebra affordances not only are important to construct a dynamic model of the task; but also offer learners and opportunity to explore relations that emerge from moving objects within the model. As a result, learners can rely on different concepts and strategies to solve the tasks. The idea in presenting this rhombus task is to illustrate that the use of a Dynamic Geometry System provides affordances for learners to construct dynamic representation of

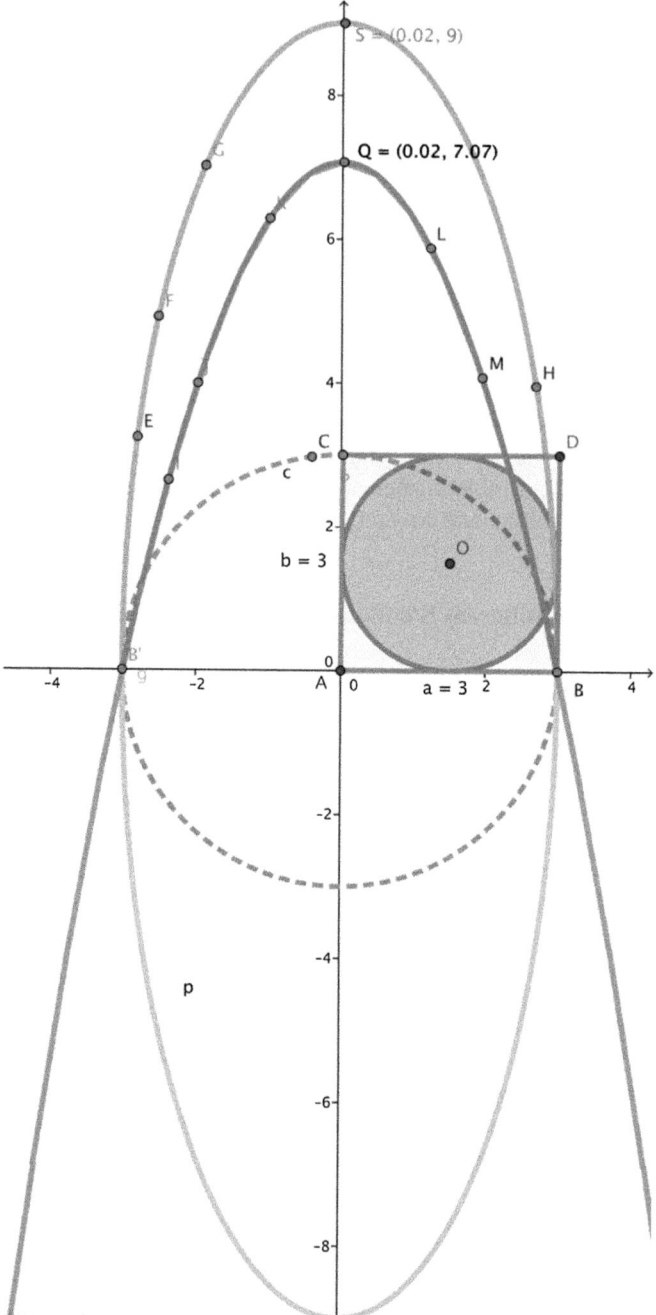

Fig. 5 Drawing the conic section that passes through five points

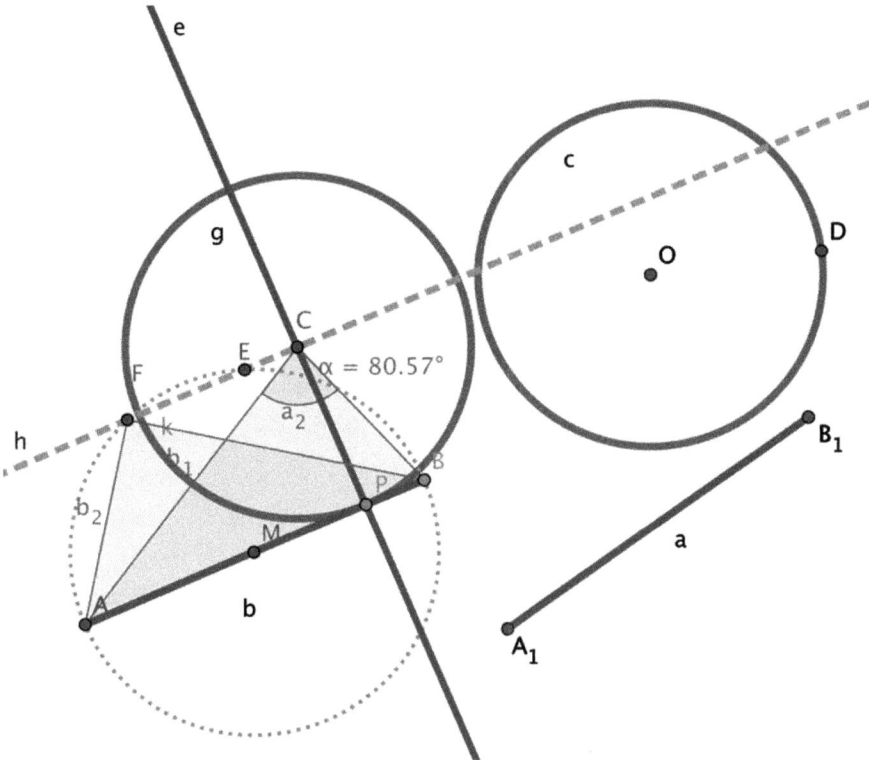

Fig. 6 Drawing segment AB tangent to the given circle

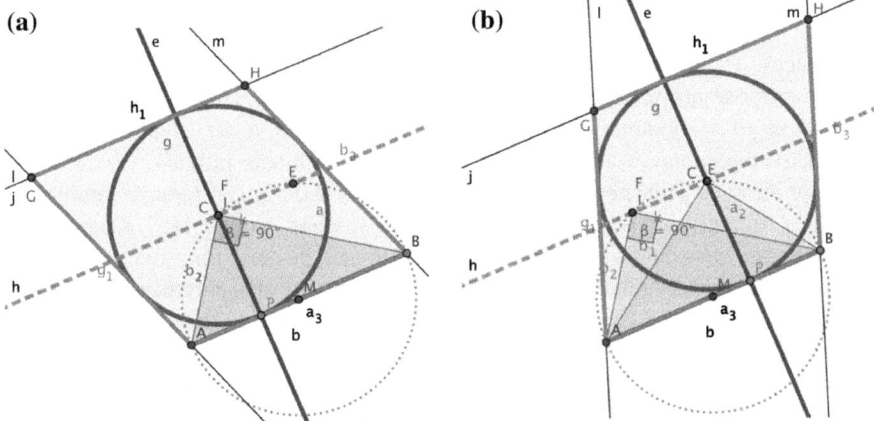

Fig. 7 a Drawing the rhombus and the inscribed circle. **b** Drawing the second solution

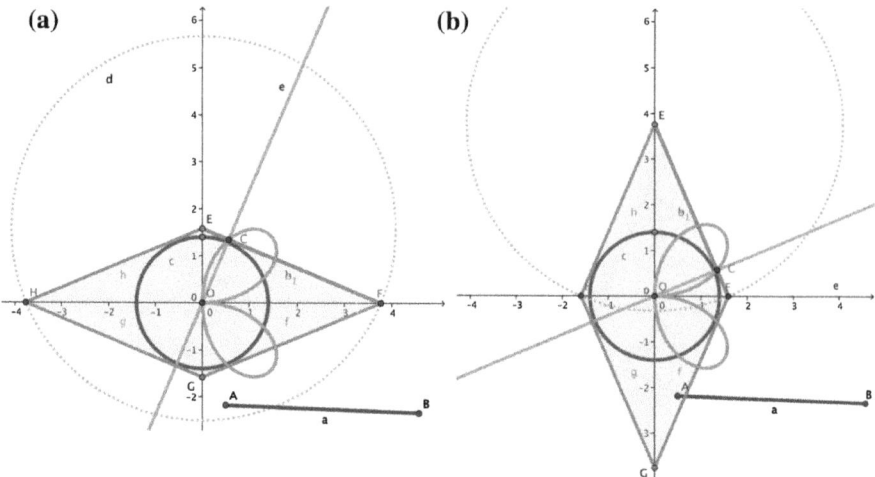

Fig. 8 **a** and **b** Another solution that involves finding a locus of point C

mathematical objects or problems, to move elements within the representation to pose questions or conjectures to explain invariants or patterns among involved parameters; to search for arguments to support emerging conjectures, and to develop a proper language to communicate results.

1.3.8 Looking Back

Conceptual frameworks used to explain learners' construction of mathematical knowledge need to capture or take into account the different ways of reasoning that students might develop as a result of using a set of tools during the learning experiences. Figure 9 show some digital technologies that learners can use for specific purpose at the different stages of problem solving activities.

The use of a dynamic system (GeoGebra) provides a set of affordances for learners to conceptualize and represent mathematical objects and tasks dynamically. In this process, affordances such as moving objects orderly (dragging), finding loci of objects, quantifying objects attributes (lengths, areas, angles, etc.), using sliders to vary parameters, and examining family of objects became important to look for invariance or objects relationships. Likewise, analysing the parameters or objects behaviours within the configuration might lead learners to identify properties to support emerging mathematical relations. Thus, with the use of the tool, learners might conceptualize mathematical tasks as an opportunity for them to engage in mathematical activities that include constructing dynamic models of tasks, formulating conjectures, and always looking for different arguments to support them. Similarly, learners can use an online platform to share their ideas, problem solutions

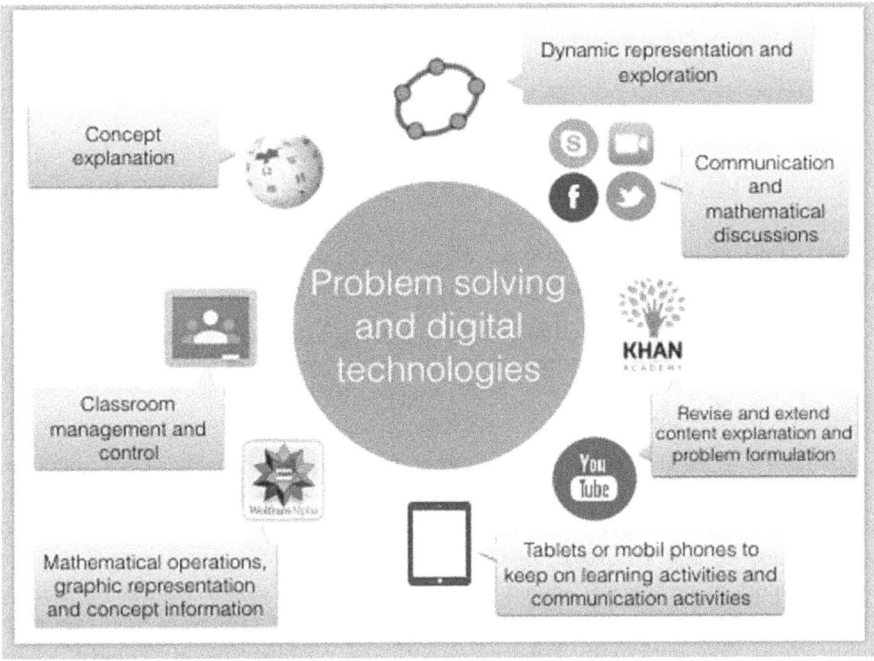

Fig. 9 The coordinated use of digital tools to engage learners in problem solving experiences

or questions in a digital wall and others students can also share ideas or solution methods and engaged in mathematical discussions that extend mathematical classroom activities.

1.4   Problem Posing: An Overview for Further Progress—Uldarico Malaspina Jurado

Problem posing and problem solving are two essential aspects of the mathematical activity; however, researchers in mathematics education have not emphasized their attention on problem posing as much as problem solving. In that sense, due to its importance in the development of mathematical thinking in students since the first grades, we agree with Ellerton's statement (2013): "for too long, successful problem solving has been lauded as the goal; the time has come for problem posing to be given a prominent but natural place in mathematics curricula and classrooms" (pp. 100–101); and due to its importance in teacher training, with Abu-Elwan's statement (1999):

While teacher educators generally recognize that prospective teachers require guidance in mastering the ability to confront and solve problems, what is often overlooked is the critical fact that, as teachers, they must be able to go beyond the role as problem solvers. That is, in order to promote a classroom situation where creative problem solving is the central focus, the practitioner must become skillful in discovering and correctly posing problems that need solutions. (p. 1)

Scientists like Einstein and Infeld (1938), recognized not only for their notable contributions in the fields they worked, but also for their reflections on the scientific activity, pointed out the importance of problem posing; thus it is worthwhile to highlight their statement once again:

The formulation of a problem is often more essential than its solution, which may be merely a matter of mathematical or experimental skills. To raise new questions, new possibilities, to regard old questions from a new angle, requires creative imagination and marks real advance in science. (p. 92)

Certainly, it is also relevant to remember mathematician Halmos's statement (1980): "I do believe that problems are the heart of mathematics, and I hope that as teachers (...) we will train our students to be better problem posers and problem solvers than we are" (p. 524).

An important number of researchers in mathematics education has focused on the importance of problem posing, and we currently have numerous, very important publications that deal with different aspects of problem posing related to the mathematics education of students in all educational levels and to teacher training.

1.4.1 A Retrospective Look

Kilpatrick (1987) marked a historical milestone in research related to problem posing and points out that "problem formulating should be viewed not only as a goal of instruction but also as a means of instruction" (Kilpatrick 1987, p. 123); and he also emphasizes that, as part of students' education, all of them should be given opportunities to live the experience of discovering and posing their own problems. Drawing attention to the few systematic studies on problem posing performed until then, Kilpatrick contributes defining some aspects that required studying and investigating as steps prior to a theoretical building, though he warns, "attempts to teach problem-formulating skills, of course, need not await a theory" (p. 124).

Kilpatrick refers to the "Source of problems" and points out how virtually all problems students solve have been posed by another person; however, in real life "many problems, if not most, must be created or discovered by the solver, who gives the problem an initial formulation" (p. 124). He also points out that problems are reformulated as they are being solved, and he relates this to investigation, reminding us what Davis (1985) states that, "what typically happens in a prolonged investigation is that problem formulation and problem solution go hand in hand, each eliciting the other as the investigation progresses" (p. 23). He also relates it to the experiences of software designers, who formulate an appropriate sequence of sub-problems to solve a problem. He poses that a subject to be examined by teachers and researchers "is whether, by drawing students' attention to the reformulating process and given them practice in it, we can improve their problem solving performance" (p. 130). He also points out that problems may be a mathematical formulation as a result of exploring a situation and, in that sense, "school exercises in constructing mathematical models of a situation presented by the teacher are intended to provide students with experiences in formulating problems." (p. 131).

Another important section of Kilpatrick's work (1987) is *Processes of Problem Formulating*, in which he considers association, analogy, generalization and contradiction. He believes the use of concept maps to represent concept organization, as cognitive scientists Novak and Gowin suggest, might help to comprehend such concepts, stimulate creative thinking about them, and complement the ideas Brown and Walter (1983) give for problem posing by association. Further, in the section "Understanding and developing problem formulating abilities", he poses several questions, which have not been completely answered yet, like "Perhaps the central issue from the point of view of cognitive science is what happens when someone formulates the problem? (…) What is the relation between problem formulating, problem solving and structured knowledge base? How rich a knowledge base is needed for problem formulating? (…) How does experience in problem formulating add to knowledge base? (…) What metacognitive processes are needed for problem formulating?"

It is interesting to realize that some of these questions are among the unanswered questions proposed and analyzed by Cai et al. (2015) in Chap. 1 of the book Mathematical Problem Posing (Singer et al. 2015). It is worth stressing the emphasis on the need to know the cognitive processes in problem posing, an aspect that Kilpatrick had already posed in 1987, as we just saw.

1.4.2 Researches and Didactic Experiences

Currently, there are a great number of publications related to problem posing, many of which are research and didactic experiences that gather the questions posed by Kilpatrick, which we just commented. Others came up naturally as reflections raised in the framework of problem solving, facing the natural requirement of having appropriate problems to use results and suggestions of researches on problem solving, or as a response to a thoughtful attitude not to resign to solving and asking students to solve problems that are always created by others. Why not learn and teach mathematics posing one's own problems?

1.4.3 New Directions of Research

Singer et al. (2013) provides a broad view about problem posing that links problem posing experiences to general mathematics education; to the development of abilities, attitudes and creativity; and also to its interrelation with problem solving, and studies on when and how problem-solving sessions should take place. Likewise, it provides information about research done regarding ways to pose new problems and about the need for teachers to develop abilities to handle complex situations in problem posing contexts.

Singer et al. (2013) identify new directions in problem posing research that go from problem-posing task design to the development of problem-posing frameworks to structure and guide teachers and students' problem posing experiences. In a chapter of this book, Leikin refers to three different types of problem posing activities, associated with school mathematics research: (a) problem posing through proving; (b) problem posing for investigation; and (c) problem posing through investigation. This classification becomes evident in the problems posed in a course for prospective secondary school mathematics teachers by using a dynamic geometry environment. Prospective teachers posed over 25 new problems, several of which are discussed in the article. The author considers that, by developing this type of problem posing activities, prospective mathematics teachers may pose different problems related to a geometric object, prepare more interesting lessons for their students, and thus gradually develop their mathematical competence and their creativity.

1.4.4   Final Comments

This overview, though incomplete, allows us to see a part of what problem posing experiences involve and the importance of this area in students mathematical learning. An important task is to continue reflecting on the questions posed by Kilpatrick (1987), as well as on the ones that come up in the different researches aforementioned. To continue progressing in research on problem posing and contribute to a greater consolidation of this research line, it will be really important that all mathematics educators pay more attention to problem posing, seek to integrate approaches and results, and promote joint and interdisciplinary works. As Singer et al. (2013) say, going back to Kilpatrick's proposal (1987),

> Problem posing is an old issue. What is new is the awareness that problem posing needs to pervade the education systems around the world, both as a means of instruction (…) and as an object of instruction (…) with important targets in real-life situations. (p. 5)

References

Abu-Elwan, R. (1999). The development of mathematical problem posing skills for prospective middle school teachers. In A. Rogerson (Ed.), *Proceedings of the International Conference on Mathematical Education into the 21st century: Social Challenges, Issues and Approaches*, (Vol. 2, pp. 1–8), Cairo, Egypt.

Ashcraft, M. (1989). *Human memory and cognition*. Glenview, Illinois: Scott, Foresman and Company.

Bailin, S. (1994). *Achieving extraordinary ends: An essay on creativity*. Norwood, NJ: Ablex Publishing Corporation.

Bibby, T. (2002). Creativity and logic in primary-school mathematics: A view from the classroom. *For the Learning of Mathematics, 22*(3), 10–13.

Brown, S., & Walter, M. (1983). *The art of problem posing*. Philadelphia: Franklin Institute Press.

Bruder, R. (2000). Akzentuierte Aufgaben und heuristische Erfahrungen. In W. Herget & L. Flade (Eds.), *Mathematik lehren und lernen nach TIMSS. Anregungen für die Sekundarstufen* (pp. 69–78). Berlin: Volk und Wissen.

Bruder, R. (2005). Ein aufgabenbasiertes anwendungsorientiertes Konzept für einen nachhaltigen Mathematikunterricht—am Beispiel des Themas "Mittelwerte". In G. Kaiser & H. W. Henn (Eds.), *Mathematikunterricht im Spannungsfeld von Evolution und Evaluation* (pp. 241–250). Hildesheim, Berlin: Franzbecker.

Bruder, R., & Collet, C. (2011). *Problemlösen lernen im Mathematikunterricht*. Berlin: CornelsenVerlag Scriptor.

Bruner, J. (1964). *Bruner on knowing*. Cambridge, MA: Harvard University Press.

Burton, L. (1999). Why is intuition so important to mathematicians but missing from mathematics education? *For the Learning of Mathematics, 19*(3), 27–32.

Cai, J., Hwang, S., Jiang, C., & Silber, S. (2015). Problem posing research in mathematics: Some answered and unanswered questions. In F.M. Singer, N. Ellerton, & J. Cai (Eds.), *Mathematical problem posing: From research to effective practice* (pp. 3–34). Springer.

Churchill, D., Fox, B., & King, M. (2016). Framework for designing mobile learning environments. In D. Churchill, J. Lu, T. K. F. Chiu, & B. Fox (Eds.), *Mobile learning design* (pp. 20–36)., lecture notes in educational technology NY: Springer.

Collet, C. (2009). Problemlösekompetenzen in Verbindung mit Selbstregulation fördern. Wirkungsanalysen von Lehrerfortbildungen. In G. Krummheuer, & A. Heinze (Eds.), *Empirische Studien zur Didaktik der Mathematik*, Band 2, Münster: Waxmann.

Collet, C., & Bruder, R. (2008). Longterm-study of an intervention in the learning of problem-solving in connection with self-regulation. In O. Figueras, J. L. Cortina, S. Alatorre, T. Rojano, & A. Sepúlveda (Eds.), *Proceedings of the Joint Meeting of PME 32 and PME-NA XXX*, (Vol. 2, pp. 353–360).

Csíkszentmihályi, M. (1996). *Creativity: Flow and the psychology of discovery and invention*. New York: Harper Perennial.

Davis, P. J. (1985). What do I know? A study of mathematical self-awareness. *College Mathematics Journal, 16*(1), 22–41.

Dewey, J. (1933). *How we think*. Boston, MA: D.C. Heath and Company.

Dewey, J. (1938). *Logic: The theory of inquiry*. New York, NY: Henry Holt and Company.

Einstein, A., & Infeld, L. (1938). *The evolution of physics*. New York: Simon and Schuster.

Ellerton, N. (2013). Engaging pre-service middle-school teacher-education students in mathematical problem posing: Development of an active learning framework. *Educational Studies in Math, 83*(1), 87–101.

Engel, A. (1998). *Problem-solving strategies*. New York, Berlin und Heidelberg: Springer.

English, L. (1997). Children's reasoning processes in classifying and solving comparison word problems. In L. D. English (Ed.), *Mathematical reasoning: Analogies, metaphors, and images* (pp. 191–220). Mahwah, NJ: Lawrence Erlbaum Associates Inc.

English, L. (1998). Reasoning by analogy in solving comparison problems. *Mathematical Cognition, 4*(2), 125–146.

English, L. D. & Gainsburg, J. (2016). Problem solving in a 21st- Century mathematics education. In L. D. English & D. Kirshner (Eds.), *Handbook of international research in mathematics education* (pp. 313–335). NY: Routledge.

Ghiselin, B. (1952). *The creative process: Reflections on invention in the arts and sciences.* Berkeley, CA: University of California Press.

Hadamard, J. (1945). *The psychology of invention in the mathematical field.* New York, NY: Dover Publications.

Halmos, P. (1980). The heart of mathematics. *American Mathematical Monthly, 87,* 519–524.

Halmos, P. R. (1994). What is teaching? *The American Mathematical Monthly, 101*(9), 848–854.

Hoyles, C., & Lagrange, J.-B. (Eds.). (2010). *Mathematics education and technology–Rethinking the terrain. The 17th ICMI Study.* NY: Springer.

Kilpatrick, J. (1985). A retrospective account of the past 25 years of research on teaching mathematical problem solving. In E. Silver (Ed.), *Teaching and learning mathematical problem solving: Multiple research perspectives* (pp. 1–15). Hillsdale, New Jersey: Lawrence Erlbaum.

Kilpatrick, J. (1987). Problem formulating: Where do good problem come from? In A. H. Schoenfeld (Ed.), *Cognitive science and mathematics education* (pp. 123–147). Hillsdale, NJ: Erlbaum.

Kline, M. (1972). *Mathematical thought from ancient to modern times.* NY: Oxford University Press.

Kneller, G. (1965). *The art and science of creativity.* New York, NY: Holt, Reinhart, and Winstone Inc.

Koestler, A. (1964). *The act of creation.* New York, NY: The Macmillan Company.

König, H. (1984). *Heuristik beim Lösen problemhafter Aufgaben aus dem außerunterrichtlichen Bereich.* Technische Hochschule Chemnitz, Sektion Mathematik.

Kretschmer, I. F. (1983). *Problemlösendes Denken im Unterricht. Lehrmethoden und Lernerfolge.* Dissertation. Frankfurt a. M.: Peter Lang.

Krulik, S. A., & Reys, R. E. (Eds.). (1980). *Problem solving in school mathematics. Yearbook of the national council of teachers of mathematics.* Reston VA: NCTM.

Krutestkii, V. A. (1976). *The psychology of mathematical abilities in school children.* University of Chicago Press.

Lesh, R., & Zawojewski, J. S. (2007). Problem solving and modeling. In F. K. Lester, Jr. (Ed.), *The second handbook of research on mathematics teaching and learning* (pp. 763–804). National Council of Teachers of Mathematics, Charlotte, NC: Information Age Publishing.

Lester, F., & Kehle, P. E. (2003). From problem solving to modeling: The evolution of thinking about research on complex mathematical activity. In R. Lesh & H. Doerr (Eds.), *Beyond constructivism: Models and modeling perspectives on mathematics problem solving, learning and teaching* (pp. 501–518). Mahwah, NJ: Lawrence Erlbaum.

Lester, F. K., Garofalo, J., & Kroll, D. (1989). *The role of metacognition in mathematical problem solving: A study of two grade seven classes.* Final report to the National Science Foundation, NSF Project No. MDR 85-50346. Bloomington: Indiana University, Mathematics Education Development Center.

Leung, A., & Bolite-Frant, J. (2015). Designing mathematical tasks: The role of tools. In A. Watson & M. Ohtani (Eds.), *Task design in mathematics education* (pp. 191–225). New York: Springer.

Liljedahl, P. (2008). *The AHA! experience: Mathematical contexts, pedagogical implications.* Saarbrücken, Germany: VDM Verlag.

Liljedahl, P., & Allan, D. (2014). Mathematical discovery. In E. Carayannis (Ed.), *Encyclopedia of creativity, invention, innovation, and entrepreneurship.* New York, NY: Springer.

Liljedahl, P., & Sriraman, B. (2006). Musings on mathematical creativity. *For the Learning of Mathematics, 26*(1), 20–23.

Lompscher, J. (1975). *Theoretische und experimentelle Untersuchungen zur Entwicklung geistiger Fähigkeiten*. Berlin: Volk und Wissen. 2. Auflage.

Lompscher, J. (1985). Die Lerntätigkeit als dominierende Tätigkeit des jüngeren Schulkindes. In L. Irrlitz, W. Jantos, E. Köster, H. Kühn, J. Lompscher, G. Matthes, & G. Witzlack (Eds.), *Persönlichkeitsentwicklung in der Lerntätigkeit*. Berlin: Volk und Wissen.

Mason, J., & Johnston-Wilder, S. (2006). *Designing and using mathematical tasks*. St. Albans: Tarquin Publications.

Mason, J., Burton, L., & Stacey, K. (1982). *Thinking mathematically*. Harlow: Pearson Prentice Hall.

Mayer, R. (1982). The psychology of mathematical problem solving. In F. K. Lester & J. Garofalo (Eds.), *Mathematical problem solving: Issues in research* (pp. 1–13). Philadelphia, PA: Franklin Institute Press.

Mevarech, Z. R., & Kramarski, B. (1997). IMPROVE: A multidimensional method for teaching mathematics in heterogeneous classrooms. *American Educational Research Journal, 34*(2), 365–394.

Mevarech, Z. R., & Kramarski, B. (2003). The effects of metacognitive training versus worked-out examples on students' mathematical reasoning. *British Journal of Educational Psychology, 73*, 449–471.

Moreno-Armella, L., & Santos-Trigo, M. (2016). The use of digital technologies in mathematical practices: Reconciling traditional and emerging approaches. In L. English & D. Kirshner (Eds.), *Handbook of international research in mathematics education* (3rd ed., pp. 595–616). New York: Taylor and Francis.

National Council of Teachers of Mathematics (NCTM). (1980). *An agenda for action*. Reston, VA: NCTM.

National Council of Teachers of Mathematics (NCTM). (2000). *Principles and standards for school mathematics*. Reston, VA: National Council of Teachers of Mathematics.

Newman, J. (2000). *The world of mathematics* (Vol. 4). New York, NY: Dover Publishing.

Novick, L. (1988). Analogical transfer, problem similarity, and expertise. *Journal of Educational Psychology: Learning, Memory, and Cognition, 14*(3), 510–520.

Novick, L. (1990). Representational transfer in problem solving. *Psychological Science, 1*(2), 128–132.

Novick, L. (1995). Some determinants of successful analogical transfer in the solution of algebra word problems. *Thinking & Reasoning, 1*(1), 5–30.

Novick, L., & Holyoak, K. (1991). Mathematical problem solving by analogy. *Journal of Experimental Psychology, 17*(3), 398–415.

Pehkonen, E. K. (1991). Developments in the understanding of problem solving. *ZDM—The International Journal on Mathematics Education, 23*(2), 46–50.

Pehkonen, E. (1997). The state-of-art in mathematical creativity. *Analysis, 97*(3), 63–67.

Perels, F., Schmitz, B., & Bruder, R. (2005). Lernstrategien zur Förderung von mathematischer Problemlösekompetenz. In C. Artelt & B. Moschner (Eds.), *Lernstrategien und Metakognition. Implikationen für Forschung und Praxis* (pp. 153–174). Waxmann education.

Perkins, D. (2000). *Archimedes' bathtub: The art of breakthrough thinking*. New York, NY: W.W. Norton and Company.

Poincaré, H. (1952). *Science and method*. New York, NY: Dover Publications Inc.

Pólya, G. (1945). *How to solve It*. Princeton NJ: Princeton University.

Pólya, G. (1949). *How to solve It*. Princeton NJ: Princeton University.

Pólya, G. (1954). *Mathematics and plausible reasoning*. Princeton: Princeton University Press.

Pólya, G. (1964). Die Heuristik. Versuch einer vernünftigen Zielsetzung. *Der Mathematikunterricht, X*(1), 5–15.

Pólya, G. (1965). *Mathematical discovery: On understanding, learning and teaching problem solving* (Vol. 2). New York, NY: Wiley.

Resnick, L., & Glaser, R. (1976). Problem solving and intelligence. In L. B. Resnick (Ed.), *The nature of intelligence* (pp. 230–295). Hillsdale, NJ: Lawrence Erlbaum Associates.

Rusbult, C. (2000). *An introduction to design*. http://www.asa3.org/ASA/education/think/intro.htm#process. Accessed January 10, 2016.

Santos-Trigo, M. (2007). Mathematical problem solving: An evolving research and practice domain. *ZDM—The International Journal on Mathematics Education, 39*(5, 6): 523–536.

Santos-Trigo, M. (2014). Problem solving in mathematics education. In S. Lerman (Ed.), *Encyclopedia of mathematics education* (pp. 496–501). New York: Springer.

Schmidt, E., & Cohen, J. (2013). *The new digital age. Reshaping the future of people nations and business.* NY: Alfred A. Knopf.

Schoenfeld, A. H. (1979). Explicit heuristic training as a variable in problem-solving performance. *Journal for Research in Mathematics Education, 10,* 173–187.

Schoenfeld, A. H. (1982). Some thoughts on problem-solving research and mathematics education. In F. K. Lester & J. Garofalo (Eds.), *Mathematical problem solving: Issues in research* (pp. 27–37). Philadelphia: Franklin Institute Press.

Schoenfeld, A. H. (1985). *Mathematical problem solving.* Orlando, Florida: Academic Press Inc.

Schoenfeld, A. H. (1987). What's all the fuss about metacognition? In A. H. Schoenfeld (Ed.), *Cognitive science and mathematics education* (pp. 189–215). Hillsdale, NJ: Lawrence Erlbaum Associates.

Schoenfeld, A. H. (1992). Learning to think mathematically: Problem solving, metacognition, and sense making in mathematics. In D. A. Grouws (Ed.), *Handbook of research on mathematics teaching and learning* (pp. 334–370). New York, NY: Simon and Schuster.

Schön, D. (1987). *Educating the reflective practitioner.* San Fransisco, CA: Jossey-Bass Publishers.

Sewerin, H. (1979): *Mathematische Schülerwettbewerbe: Beschreibungen, Analysen, Aufgaben, Trainingsmethoden mit Ergebnissen.* Umfrage zum Bundeswettbewerb Mathematik. München: Manz.

Silver, E. (1982). Knowledge organization and mathematical problem solving. In F. K. Lester & J. Garofalo (Eds.), *Mathematical problem solving: Issues in research* (pp. 15–25). Philadelphia: Franklin Institute Press.

Singer, F., Ellerton, N., & Cai, J. (2013). Problem posing research in mathematics education: New questions and directions. *Educational Studies in Mathematics, 83*(1), 9–26.

Singer, F. M., Ellerton, N. F., & Cai, J. (Eds.). (2015). *Mathematical problem posing. From research to practice.* NY: Springer.

Törner, G., Schoenfeld, A. H., & Reiss, K. M. (2007). Problem solving around the world: Summing up the state of the art. *ZDM—The International Journal on Mathematics Education, 39*(1), 5–6.

Verschaffel, L., de Corte, E., Lasure, S., van Vaerenbergh, G., Bogaerts, H., & Ratinckx, E. (1999). Learning to solve mathematical application problems: A design experiment with fifth graders. *Mathematical Thinking and Learning, 1*(3), 195–229.

Wallas, G. (1926). *The art of thought.* New York: Harcourt Brace.

Watson, A., & Ohtani, M. (2015). Themes and issues in mathematics education concerning task design: Editorial introduction. In A. Watson & M. Ohtani (Eds.), *Task design in mathematics education, an ICMI Study 22* (pp. 3–15). NY: Springer.

Zimmermann, B. (1983). Problemlösen als eine Leitidee für den Mathematikunterricht. Ein Bericht über neuere amerikanische Beiträge. *Der Mathematikunterricht, 3*(1), 5–45.

Further Reading

Boaler, J. (1997). *Experiencing school mathematics: Teaching styles, sex, and setting.* Buckingham, PA: Open University Press.

Borwein, P., Liljedahl, P., & Zhai, H. (2014). Mathematicians on creativity. Mathematical Association of America.

Burton, L. (1984). *Thinking things through.* London, UK: Simon & Schuster Education.

Feynman, R. (1999). *The pleasure of finding things out.* Cambridge, MA: Perseus Publishing.

Gardner, M. (1978). *Aha! insight.* New York, NY: W. H. Freeman and Company.

Gardner, M. (1982). *Aha! gotcha: Paradoxes to puzzle and delight*. New York, NY: W. H. Freeman and Company.

Gardner, H. (1993). *Creating minds: An anatomy of creativity seen through the lives of Freud, Einstein, Picasso, Stravinsky, Eliot, Graham, and Ghandi*. New York, NY: Basic Books.

Glas, E. (2002). Klein's model of mathematical creativity. *Science & Education, 11*(1), 95–104.

Hersh, D. (1997). *What is mathematics, really?*. New York, NY: Oxford University Press.

Root-Bernstein, R., & Root-Bernstein, M. (1999). *Sparks of genius: The thirteen thinking tools of the world's most creative people*. Boston, MA: Houghton Mifflin Company.

Zeitz, P. (2006). *The art and craft of problem solving*. New York, NY: Willey.

www.ingramcontent.com/pod-product-compliance
Ingram Content Group UK Ltd.
Pitfield, Milton Keynes, MK11 3LW, UK
UKHW020217231225
466357UK00011B/182